Alchemy And The Alembic

STEVEN SCHOOL

DISCLAIMER

This book is intended for informational purposes only, neither the author nor the publisher assume any liability for the use or misuse of the information provided here. No warranty is expressed or implied as to the accuracy or completeness of any information contained in this book.

CONTENTS

ACKNOWLEDGMENTS

I would like to acknowledge Nicholas and Perenelle Flamel,
Michael Sendivogius, Alexander Seton,
Theophrastus Paracelsus, Friederich Gualdus,
Edward Kelly, John Dee, Hermes Trismegistus,
Zosimos of Panopolis, Pliny the Elder, John Pontanus,
Abraham Eleazer, and Maria the Prophetess.

1 INTRODUCTION

The Alembic was a medieval distillation vessel that was an integral part of the hermetic art of primitive Alchemy. In this book we will discuss this indispensable utensil as well as its basic purpose. We will also include pictures of my own modern version of an Alembic which I assembled and use myself quite efficiently in my alchemical works. This apparatus consists of a lower portion which resembles a modern round boiling flask, which was called the curcurbit. The matter to be distilled was placed here, this vessel was also used for the putrefaction process which is also called fermentation. For this specific purpose a blind head was sometimes mounted on top of the curcurbit which resembles an inverted round boiling flask, it is similar to a distillation head, except for the fact that it has no outlets, therefore everything that distills upward, eventually coagulates and returns to the lower portion of the glass vessel. The next item of importance is the distillation head itself, which in antiquity had a long arm attached to it which served as a condenser, followed by the receptacle which would then be attached to the end of that arm.

Both the Alembic itself and its accessories were made of glass, the parts were fashioned by primitive glass blowers, there are craftsmen in some parts of the world who can still make these today. Some of the key points to look for when purchasing an Alembic are, the thickness and quality of the glass, is the material rated to withstand heat?, does the distillation head have a rain gutter formed into it that will direct everything that rises and coagulates into the receptacle so that it cannot drip back down into the curcurbit?, does your alembic come with both a blind head and a distillation head?, is there a warranty against breakage?, is the end of the distillation condenser arm fitted to receive a ground glass receptacle? And last but not least, what is the size of the opening at the connection point between the upper and lower portions of the apparatus?, the larger the opening the better, even if it has to be custom made, because when the volatile salts crystallize in the distillation head, it can be very frustrating if the opening is too small to efficiently retrieve them. If the opening is large enough that you can reach your hand inside with a plastic spoon, that is excellent, if not, at least get the opening constructed as large as possible. My own version of an alembic is constructed of modern borosilicate distillation vessels which can be superior to its medieval counterpart, since it can be further taken apart, as well as configured in different ways. The openings on mine however are only 24/40. A larger size of 45/50 would allow easier access without having to get it custom made. The alembic was also used for another alchemical process which was known as circulation.

the alchemists could use their alembic for the lower circulation which pertains to the plant realm, and is called spagyrics, this involves the creation of various remedies and elixirs which are derived from plants, medieval medicines so to speak. My own work and alchemical studies however mainly pertain to the mineral kingdom and in this area the sages used their alembic for the creation of what was called the alchahest, it is also sometimes referred to in alchemical tracts as a work of Saturn, our old one, antimony, our antimonial compound consisting of saturnine parts, or the voracious wolf. The old sages sometimes said of this mysterious substance that you must give it gold, or another substance to work upon, lest it would begin to dissolve the glass that it was contained in. and this then, is properly known as the urine work which is among the items discussed and clarified in Alchemy And The Vinegar Of The Sages DVD available at

http://www.howtomakethephilosophersstone.com

And now we will advance to pictures concerning the basic construction and use of my own modern alembic. This book is intended for informational purposes only, do not try this at home, and do not ingest any alchemical substances of any sort. My work is not intended to treat or cure any disease.

Modern Alembic made from boiling flasks and related fittings.

I have the receptacle slightly vented to prevent pressure from building up within the glass for this work. The distillate here will condense in the distillation head and then drip down into the receptacle, this method is effective and has allowed the volatile salt to crystallize within the inverted two neck boiling flask which can be easily removed and sealed with ground glass stoppers until I am ready to harvest the salt. The receptacle can also be easily removed and stoppered to save the distillate. The caput mortuary will be left in the curcurbit, it will be calcined at a later time and then leached to extract the fixed salt.

The salt volatile.

The distillation head. A blind head can be installed by simply utilizing a round boiling flask with only one neck.

The curcurbit.

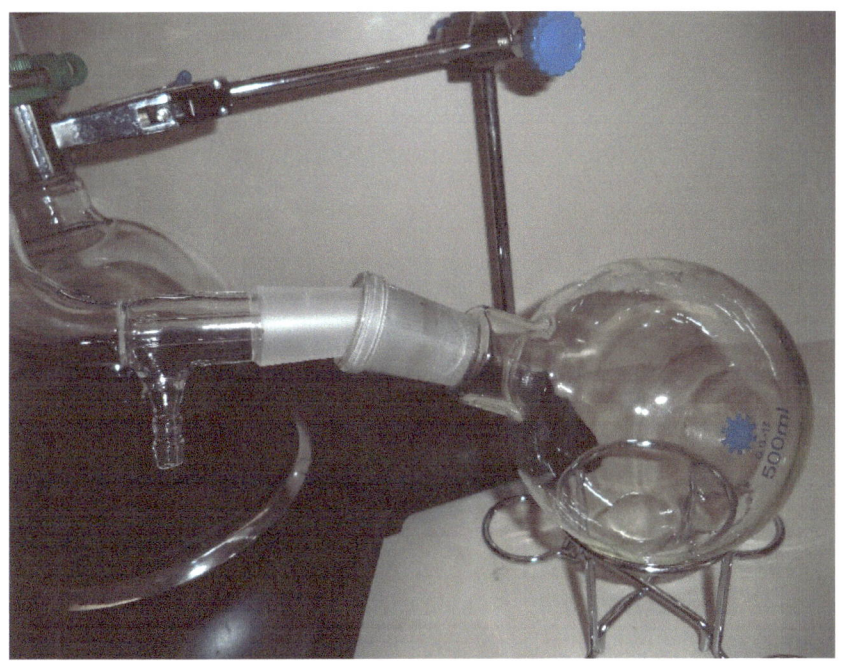

The receptacle.

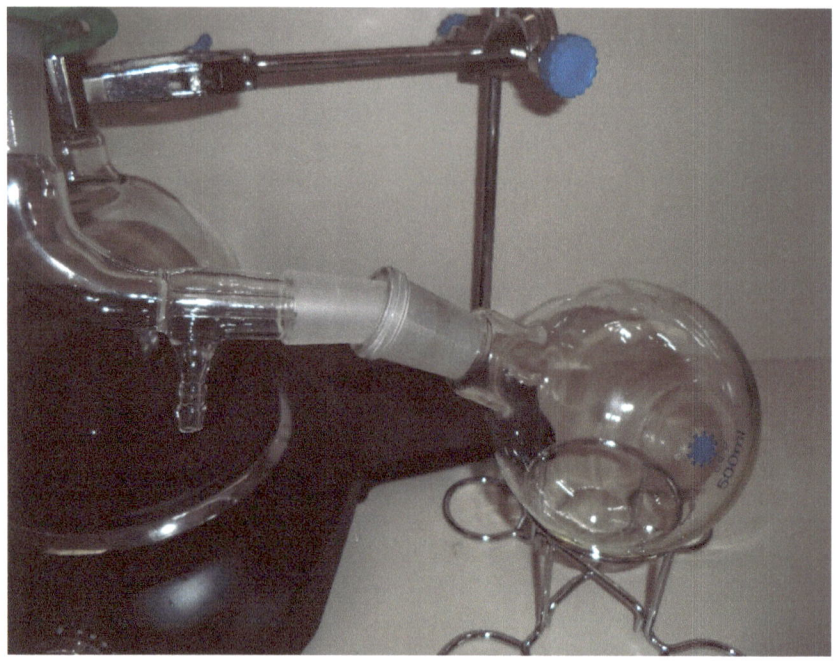

The receptacle again, slightly different angle. Notice the clear distillate is forming a puddle.

The salt is as clear as liquid crystal. (Volatile salt of putrefied urine shown here.)

Calcination of the caput mortuary after distillation is completed. (Done outdoors).

Extraction of the caput mortuary, leaching out the fixed salts.

Filtering the liquid to separate the black matter/impurity.

Evaporation of the filtered liquid to obtain the fixed salt which will need further purification, as you can see it is already beginning to precipitate however it still has some blackness in it.

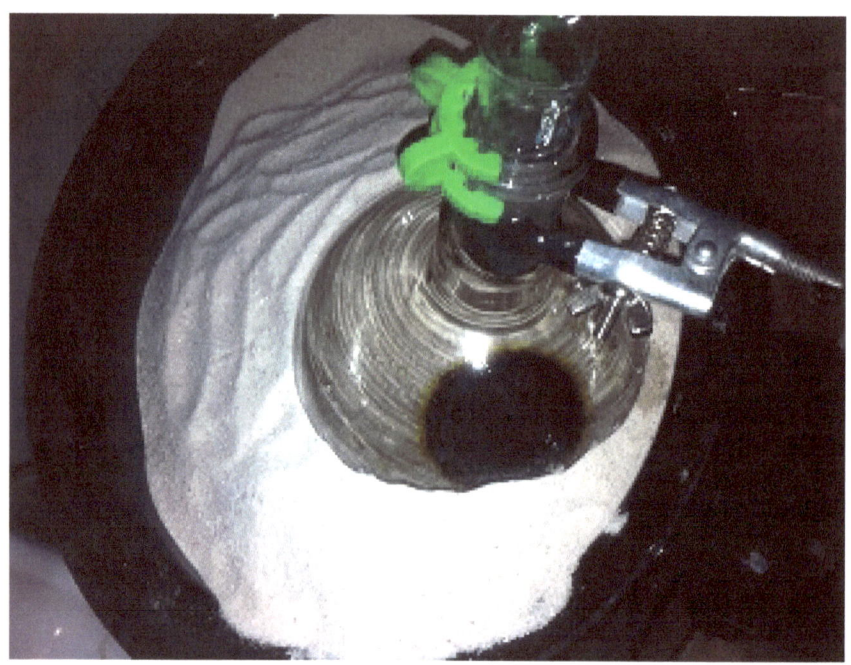

Putrefaction in sand bath.

Alembic with blind head.

WWW.HOWTOMAKETHEPHILOSOPHERSSTONE.COM

Work of Steven School.

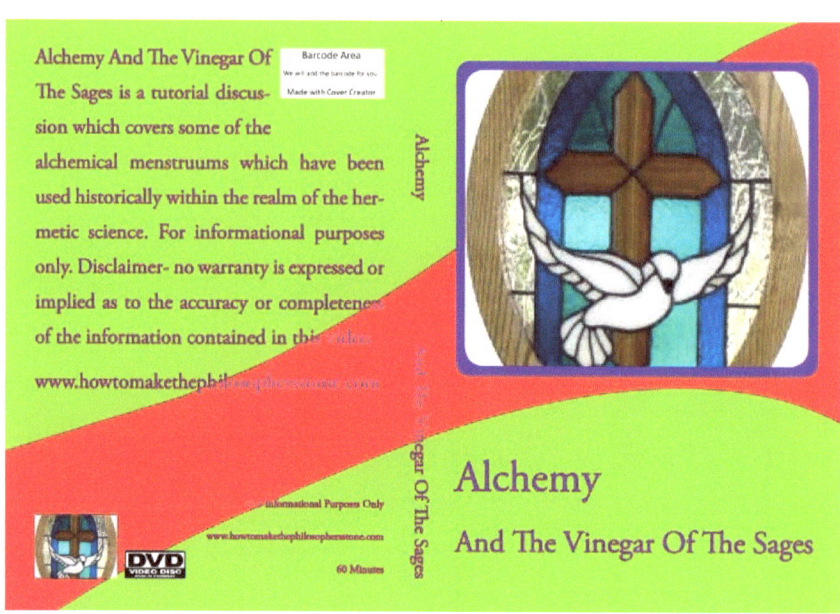

Alchemy And The Vinegar Of The Sages is a tutorial discussion which covers some of the alchemical menstruums which have been used historically within the realm of the hermetic science. For informational purposes only. Disclaimer- no warranty is expressed or implied as to the accuracy or completeness of the information contained in this video

www.howtomakethephilosophersstone.com

Informational Purposes Only

www.howtomakethephilosophersstone.com

60 Minutes

DVD VIDEO DISC

Alchemy

And The Vinegar Of The Sages

http://www.howtomakethephilosophersstone.com

SOL and LUNA

In this book, we will discuss sol and luna, the salt and the living water which make up the hidden stone. we will cover the required alchemical vessels, the technique of conjunction, imbibition, multiplication, we will touch upon the veritable prima materia itself from which the stone is made, and give you a fighting chance, to uncover this great art. King and Queen, Sun and Moon. Sol and Luna. the salt and the water, the body and the spirit. philosophical gold and living silver.

————————— About the Author —————————

The author of this informational series undertook the task of performing a relentless and independent study of alchemy while following no teacher. In this booklet we will discuss and clarify Sol and Luna, we will cover processes and techniques of conjunction, imbibition, and multiplication. This booklet is the result of my own research and experiments in the ancient hermetic art of alchemy. For informational purposes only, no warranty is expressed or implied as to the accuracy or completeness of the information contained within this booklet.

http://www.howtomakethephilosophersstone.com

Barcode Area

We will add the barcode for you.

Made with Cover Creator

Steven School   **SOL AND LUNA**

THE HERMETICAL
WEDDING

www.createspace.com/4191585

27

ᘯ
alchemy and the green lion

This is the first alchemy book that i wrote, This booklet represents some of my alchemical discoveries from 4 years of hands on research and experimentation regarding alchemy and the philosophers stone, it is meant to be simple and to the point. for informational purposes only, no warranty is expressed or implied as to the accuracy or completeness of the information contained within the pages of this booklet.
http://www.howtomakethephilosophersstone.com

About the Author

The author wrote this book after four years of study and research in the mineral kingdom of alchemy with no teacher, since then Steven has made major breakthroughs in the ancient hermetic art.

Barcode Area

We will add the barcode for you.

Made with Cover Creator

alchemy and the
green lion

ᘰ

steven school

the truth of the
philosophers stone

.

ABOUT THE AUTHOR

http://www.howtomakethephilosophersstone.com

www.ingramcontent.com/pod-product-compliance
Lightning Source LLC
Chambersburg PA
CBHW041151180526
45159CB00002BB/782